AF598772

MILITARY VEHICLES

BATTLESHIPS

MARTY GITLIN

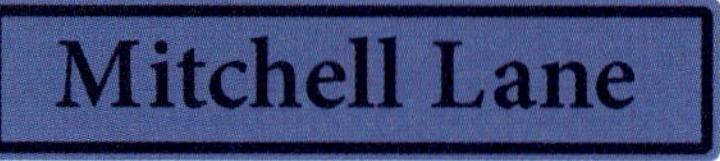

PUBLISHERS

2001 SW 31st Avenue
Hallandale, FL 33009
www.mitchelllane.com

First Edition, 2021.

Author: Marty Gitlin
Designer: Ed Morgan
Editor: Morgan Brody

Little Mitchie is an imprint of Mitchell Lane Publishers.

Title: Military Vehicles: Battleships / by Marty Gitlin
Description: Hallandale, FL :
Mitchell Lane Publishers, [2021]

Series: Military Vehicles
Library bound ISBN: 978-1-58415-240-8
eBook ISBN: 978-1-58415-254-5

Photo credits: Freepik.com, Freevector.com, United States public domain, p. 10 Dom0803 at English Wikipedia

CONTENTS

Words in **bold** can be found in the Glossary.

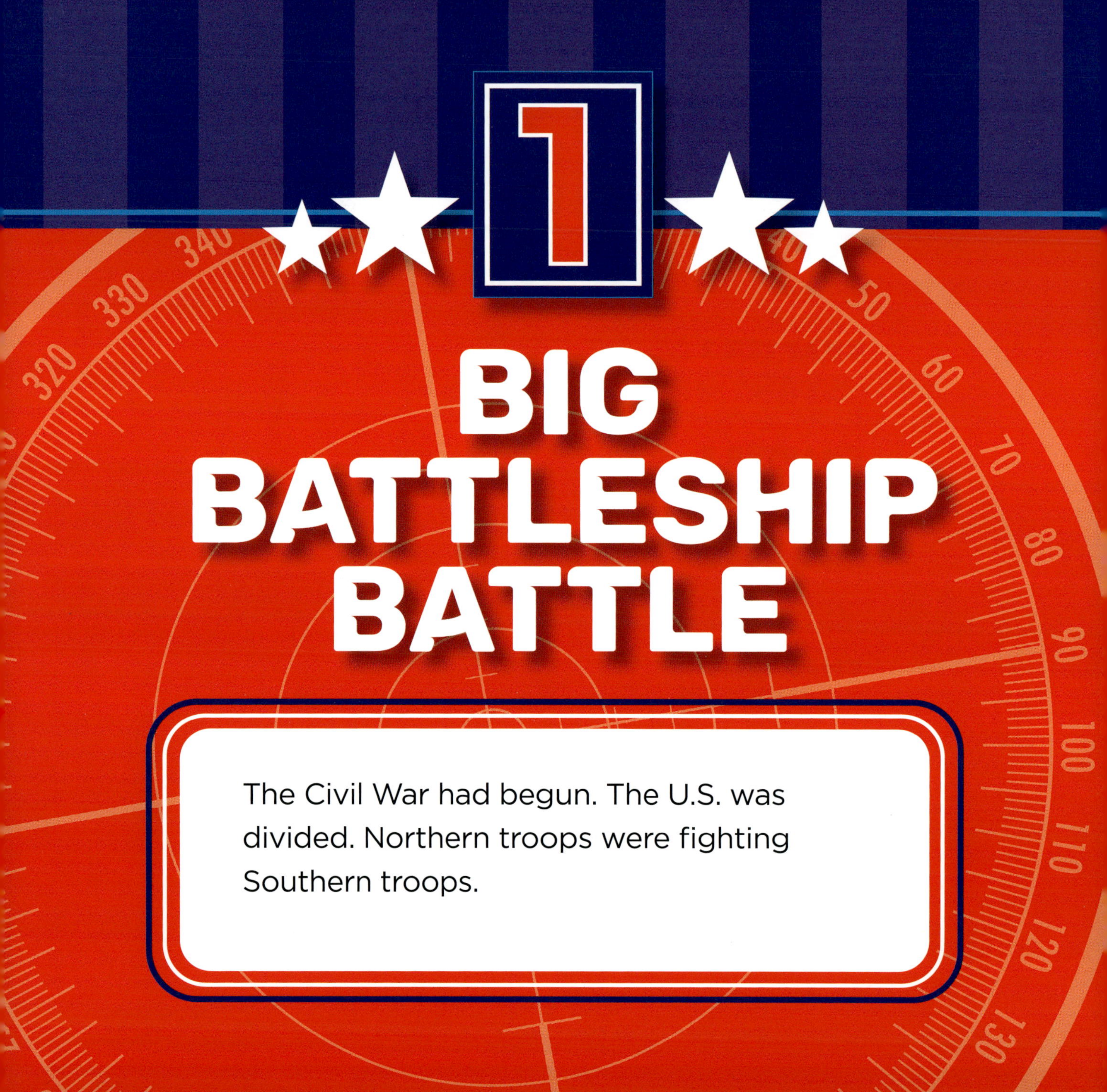

1

BIG BATTLESHIP BATTLE

The Civil War had begun. The U.S. was divided. Northern troops were fighting Southern troops.

It was March 9, 1862. Many men were dying in battle. But it was a battle of battleships that made big news that day. Those battleships were the *Monitor* and *Merrimac*.

The *Monitor* fought for the North. The *Merrimac* fought for the South. They were the first **ironclad** warships.

This is the deck of the USS *Monitor* while in the James River in Virginia in July 1862.

The two ships battled off the shores of Virginia. They shot cannon balls at each other. None did much damage. They bounced off the iron **hulls**.

Their clash had little effect on the Civil War. But they changed **naval** warfare forever. The U.S. Navy stopped building wooden warships.

The *Monitor* and *Merrimac* made an impact. They launched the era of battleships.

An artist's rendering of the battle between the *Merrimac* (*left*) and the *Monitor*.

FAST FACT

GOING DOWN WITH THE SHIP

The *Merrimac* and *Monitor* did not last past 1862. Southern troops soon destroyed the *Merrimac*. They did not want it falling into enemy hands. The *Monitor* sank that December.

The USS *Monitor* sinks in 1862.

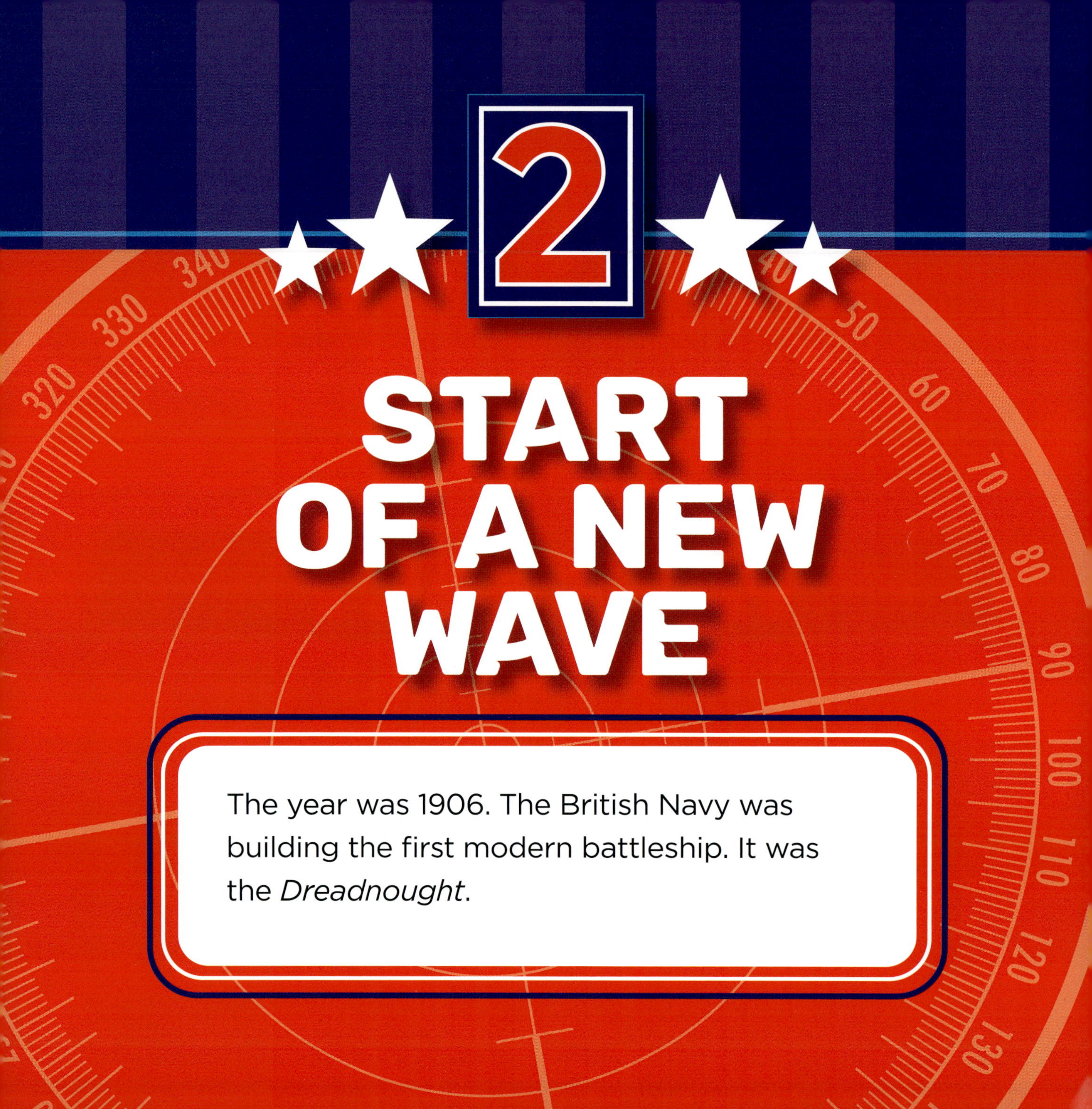

2

START OF A NEW WAVE

The year was 1906. The British Navy was building the first modern battleship. It was the *Dreadnought*.

The **vessel** was fast. It could travel a long way. Its large guns blasted from their **turrets**. They punched holes in enemy ships. The *Dreadnought* was strong. It had heavy steel **armor**.

The British Navy became the best in the world. It soon built even better battleships. They could fire **torpedoes** to sink their targets.

The British battleship HMS *Dreadnought* is shown in 1906.

Germany tried to catch up. Its **fleet** fought the British in World War I. Both nations wanted control of the North Atlantic Ocean.

Their ships clashed in the 1916 Battle of **Jutland**. Neither side won. Both countries had to keep improving their battleships. Those were built with even stronger guns and armor.

HMS *Caroline* is the last surviving warship that was at the Battle of Jutland and is preserved in Belfast, Northern Ireland.

U.S. battleship *New Jersey* in a camouflage coat in 1918.

FAST FACT

DON'T LEAVE THE FLEET!

World War I battleships could not **stray**. All their battles were fought in close range. One reason was that ship crews had to talk to each other. But radio signals did not travel far.

3

HOW THEY HELPED

The British were not alone in building warships in 1906. The U.S. Navy was doing the same. That was the year it launched its Atlantic fleet. It included the Battleship Force.

The U.S. fleet made history. They were the first warships to sail around the world. The 14-month journey ended in 1909.

The USS *Virginia* is anchored in 1909 with a USS *Illinois* class battleship in the background.

The most heavily armored U.S. battleship of the era was the *Wyoming*. It was 562 feet long. It carried more than 1,000 men. It featured 33 guns and two torpedo tubes.

World War II put American battleships to the test. They were used to **patrol** the seas and protect shipping lanes.

The USS Wyoming is shown around 1908.

The USS *Maine*.

FAST FACT ABOUT THE *MAINE*

The first U.S. battleship was the *Maine*. It was built in 1888. It cost more than $2 million to build. Today it would cost over $26 million. An explosion in 1898 killed 268 of its crewmen.

4

WHY DID THEY WIN?

Battleships are no longer used in warfare. But they once served a great purpose.

Those U.S. ships helped win World War II. They had many uses. Some **escorted** other ships. Others defended air attacks. Others used their guns to support ground troops near the shores.

The ships of that era were much bigger and stronger. They weighed up to 120 million pounds.

The USS *Missouri* (*left*) transfers sailors and officers to the USS *Iowa* during the Japanese surrender in 1945.

Their guns were also far more powerful. Those on the *Maine* were 10 inches long. They fired about 3,000 yards. The battleship guns of World War II were 16 inches. They reached targets 20 miles away!

An American battleship deck during World War II.

FAST FACT

DOING THE JOB ON D-DAY

U.S. battleships played a huge role on D-Day. The date was June 6, 1944. American troops helped invade France that day. Germany had taken over that country. Many U.S. battleships fired at the Germans on the shores of France. They gave cover to the troops. That saved many lives.

Soldiers wade ashore on Omaha Beach on D-Day.

5

NO LONGER NEEDED

Why are battleships history? They no longer serve a purpose. They were too heavy and slow.

The U.S. Navy needed their craft to do other tasks. They used aircraft carriers in World War II and beyond. Their planes could take off from the deck and bomb targets. Battleships could not carry warplanes.

American battleships did stay useful for a while. They fired **shells** to support ground troops during Operation **Desert Storm** in the 1990s. But the Navy has not used battleships since 1992.

FAST FACT

A TRIP TO THE MUSEUM

The most famous World War II battleships are in museums. Each was named after a state. They are the *Missouri*, *Iowa*, *New Jersey*, and *Wisconsin*.

What You Should Know

- Battleships have been protected over the years by wood, iron and steel.
- The battle between the *Monitor* and *Merrimac* was a highlight of the Civil War.
- The height of battleship warfare was in World War I.
- U.S. battleships have ranged from 20 million to 120 million pounds.
- Battleship guns could fire as far as 20 miles.
- The primary use of World War II battleships was to support ground troops on shore.
- The Japanese destroyed two battleships in their Pearl Harbor attack.
- American battleships have not been used in warfare since 1992.

Glossary

armor
Metal covering to protect ships

Desert Storm
Early 1990s war against Iraq after its invasion of Kuwait

escort
To accompany another ship for protection

fleet
A group of military ships

hull
Main body of a ship

ironclad
Covered or protected with iron

Jutland
Peninsula of north-western Europe

naval
Relating to a navy or navies

patrol
Group of people that cover an area to watch over it

shell
An explosive warhead or bomb

stray
To go in a direction that is away from a group

torpedo
A propelled missile fired from a ship

turret
Revolving tower with guns that shoot from a warship

vessel
A ship or large boat

Find Out More

Drennan Demuth, Patricia. *What was D-Day?* East Rutherford, N.J.: Penguin Publishing, 2015.

Zuehlke, Jeffrey. *Warships* (Pull Ahead Books - Mighty Movers). Minneapolis, MN: First Avenue Editions, 2005.

Batchelor, John. *World War II Warships*. Mineola, N.Y.: Dover Publications, 2006.

Websites

Ducksters: This site gives insight into the Civil War battle between the *Monitor* and *Merrimac*.
https://www.ducksters.com/history/civil_war/battle_of_ironclads.php

Britannica Kids (Navy): Find out all about the U.S. Navy on this website.
https://kids.britannica.com/kids/article/navy/353522

National Park Service: This site provides information and photos of famous U.S. Navy battleships.
https://www.nps.gov/valr/learn/historyculture/battleship-row.htm

Index

About the Author

Martin Gitlin is a freelance writer based in Cleveland, Ohio. He has authored about 150 books since 2006. The majority were written for young students. He boasts an interest in military history and strong knowledge of battleships. Martin served as a newspaper journalist from 1991 to 2002. He earned more than 45 awards during that time. Included was a first place for general excellence from Associated Press in 1996. That organization also voted him as one of the top four feature writers in Ohio in 2002.